Environment *in Focus*

Pollution

Cheryl Jakab

Marshall Cavendish
Benchmark
New York

Library of Congress Cataloging-in-Publication Data

Jakab, Cheryl.
   Pollution / Cheryl Jakab.
      p. cm. — (Environment in focus)
   Summary: "Discusses the environmental issue of pollution and how to create
   a sustainable way of living"—Provided by publisher.
   Includes bibliographical references and index.
   ISBN 978-1-60870-089-9
   1. Pollution—Juvenile literature. I. Title.
   TD176.J348 2011
   363.73—dc22
                              2009042327

First published in 2010 by
MACMILLAN EDUCATION AUSTRALIA PTY LTD
15–19 Claremont Street, South Yarra 3141

Visit our website at www.macmillan.com.au or go directly to www.macmillanlibrary.com.au

Associated companies and representatives throughout the world.

Copyright © Cheryl Jakab 2010

Edited by Margaret Maher
Text and cover design by Cristina Neri, Canary Graphic Design
Page layout by Domenic Lauricella
Photo research by Sarah Johnson
Illustrations by Domenic Lauricella
Maps courtesy of Geo Atlas

Printed in the United States

Acknowledgments
The author and the publisher are grateful to the following for permission to reproduce copyright material:

Front cover photograph: A sea bird covered in oil, photo-illustration by Warren Hackshall/ Fairfax Photos

AAP Image/AP, 7 (bottom left), 21; Image copyright (c) Clean Up Australia, www.cleanup.org.au, 11; © Pallava Bagla/CORBIS, 20; © Bettmann/CORBIS, 22; © Construction Photography/CORBIS, 18; © Erik Schaffer; Ecoscene/CORBIS, 16; © Viviane Moos/CORBIS, 12; © Skyscan/CORBIS, 24; Fairfax Photos/Andrew De La Rue, 19; China Photos/Stringer/Getty Images, 5; Tom Paiva/Getty Images, 6 (top), 13; Masterfile/Mark Downey, 26; Image © NASA, 15; Peter Bennett/Photolibrary, 28; Robert Brooks/SPL/Photolibrary, 10; Simon Fraser/Science Photo Library/Photolibrary, 23; Andre Maslennikov/Photolibrary, 7 (top), 9; REUTERS/Steve Marcus, 7 (bottom right), 27; © foment/Shutterstock, 6 (bottom), 17; © Stephen Gibson/Shutterstock, 25; © Happy Alex/Shutterstock, 14; © Joseph McCullar/Shutterstock, 8; Image © Environmental Memoirs, National Center for Sustainability, Swinburne. Photographer Eider Huertas, 29.

While every care has been taken to trace and acknowledge copyright, the publisher tenders their apologies for any accidental infringement where copyright has proved untraceable. Where the attempt has been unsuccessful, the publisher welcomes information that would redress the situation.

Please note
At the time of printing, the Internet addresses appearing in this book were correct. Owing to the dynamic nature of the Internet, however, we cannot guarantee that all these addresses will remain correct.

1 3 5 6 4 2

Contents

Glossary Words
When a word is printed in **bold**, you can look up its meaning in the Glossary on page 31.

Environment in Focus

Hi there! This is Earth speaking. Will you spare a moment to listen to me? I have some very important things to discuss.

We must focus on some urgent environmental problems! All living things depend on my environment, but the way you humans are living at the moment, I will not be able to keep looking after you.

The issues I am worried about are:
- large ecological footprints
- damage to natural wonders
- widespread pollution in the environment
- the release of **greenhouse gases** into the **atmosphere**
- poor management of waste
- environmental damage caused by food production

My challenge to you is to find a **sustainable** way of living. Read on to find out what people around the world are doing to try to help.

Fast Fact
Concerned people in local, national, and international groups are trying to understand how our way of life causes environmental problems. This important work helps us learn how to live more sustainably now and in the future.

What's the Issue?
Pollution

Pollutants produced by human activities are now widespread across Earth. They cause many health and environmental problems.

Sources of Pollution

Sources of pollution range from people throwing away food packaging to oil spills from giant tankers. Some pollution, such as excessive noise, can simply make life unpleasant. Other pollution, such as car exhaust fumes, can be harmful to people and other living things.

Dealing with Pollution

The best way to deal with pollution is to prevent it in the first place. Existing pollution is dealt with in different ways, depending on the type of pollutant and how widespread it is. For example, plastic litter must be collected and removed. Light pollution can be dealt with by removing the source.

Fast Fact
Haze is a type of **visual pollution**. It is the gray clouds around cities when they are viewed from a distance.

Water pollution is unpleasant and can harm plants and animals that live in the water.

Pollution Issues

The most urgent environmental pollution issues around the globe include:

- widespread sources of pollution
- damage from air pollution
- build-up of litter
- pollution in waterways and oceans
- increases in noise and light pollution

ARCTIC OCEAN

Arctic Circle

NORTH

AMERICA

California

NORTH

ATLANTIC

OCEAN

Pacific Ocean

SOUTH

AMERICA

ISSUE 2

California
Trucks at Long Beach Port are causing harmful air pollution. See pages 12–15.

ISSUE 3

Pacific Ocean
A huge drifting island of plastic litter is collecting. See pages 16–19.

Fast Fact
The part of an ecological footprint that is due to greenhouse gas **emissions** is called a carbon footprint.

Around the Globe

ISSUE 1

Baltic Sea
Dead zones are being caused by widespread pollution, including fertilizer runoff and emissions from burning **fossil fuels**. See pages 8–11.

Baltic Sea

EUROPE

France

ASIA

Japan

AFRICA

NORTH

PACIFIC

Tropic of Cancer

OCEAN

Equator

INDIAN

OCEAN

ISSUE 4

France
The *Amoco Cadiz* oil spill has polluted the ocean and damaged coasts and sea life. See pages 20–23.

ISSUE 5

Japan
Light pollution is reducing people's quality of life. See pages 24–27.

Pollution from Widespread Sources

Pollution from widespread sources is reducing people's quality of life and damaging natural ecosystems.

Point and Nonpoint Sources

Pollution can come from a point source or a nonpoint source. Point-source pollution is from a single event or source, such as a fire. Nonpoint-source pollution comes from a widespread source, such as excess fertilizer running off pastures. Nonpoint-source pollution can be very difficult to control.

Unseen Pollution

Unseen pollution can be more damaging than pollution that can be seen. This is partly because unseen pollution is hard to detect and so remains untreated. For example, litter is pollution that can be seen. However, if this litter is not cleaned up it can wash into oceans. In the oceans, where the litter is unseen, it can collect to become an even bigger problem.

Point-source pollution, such as smoke from a factory fire, can be controlled once the source is identified.

Fast Fact
Some environmental pollutants are poisonous. However, most pollutants are not poisonous. They damage environments when they become widespread in large volumes.

Large algal blooms in the Baltic Sea have caused dead zones to develop.

CASE STUDY
Baltic Sea Dead Zones

Dead zones are areas of the sea with no fish or other aquatic life. The Baltic Sea has seven of the ten largest dead zones in the world. If these areas continue to increase in size, the Baltic Sea's ecosystem could collapse.

What Causes Dead Zones?

Areas in oceans become dead zones when the water becomes polluted with chemical nutrients such as fertilizer. These nutrients cause **algae** to grow into massive **algal blooms**. Eventually, these algal blooms die off, and the decaying algae use up the oxygen in the water. Most sea life cannot survive in low-oxygen conditions, so the area becomes a dead zone.

Chemical nutrients can come from **agricultural runoff**, emissions from burning fossil fuels, large amounts of human waste, and offshore fish farming.

Fast Fact

There are an estimated 146 dead zones in coastal waters worldwide. The number of dead zones has doubled each decade since the 1960s.

Toward a Sustainable Future: Pollution Awareness

Pollution awareness is not just a matter of wanting a clean local area. It involves understanding that pollution builds up over time and can affect regions distant from the source.

Regulation of Dangerous Substances

Many dangerous substances are used in industry before their damaging effects are known. Once the effects are known, dangerous and polluting substances, such as chemicals from paper mills, can be controlled through regulations. When these substances are regulated, the amounts in the environment decrease. Laws now regulate human exposure to thousands of dangerous substances.

Learning About Pollution

Learning about where pollution comes from helps people understand how they can reduce pollution. It can also help people understand how to clean up polluted areas.

Litter that is dropped in gutters can be washed into the sea when it rains.

Fast Fact
The United States Environmental Protection Agency runs a pollution prevention campaign, called P2, in September each year. This campaign highlights the need to reduce pollution from all sources.

On World Environment Day 2009 volunteers in Victoria, Australia, planted trees to help prevent climate change.

CASE STUDY
World Environment Day

World Environment Day is held on June 5 each year. It highlights the need to care for the world's environments for our long-term health.

Improving World Environments

World Environment Day helps individuals, companies, and governments to improve world environments. It was set up by the United Nations in 1972. Its purpose is to interest people in making human activities more environmentally sustainable.

World Environment Day Themes and Hosts

Since 1987, each World Environment Day has had a host country and a theme. The host country for 2009 was Mexico. The theme was "Your planet needs you! UNite to combat climate change." This theme focused attention on the urgent need to tackle climate change. The 2009 World Environment Day also addressed the need to overcome poverty and improve forests.

Fast Fact
The World Environment Day theme for 2007 was "Melting ice: a hot topic?" In 2008 the theme was "Kick the CO_2 habit: toward a low carbon economy."

11

Air Pollution

Air pollution is any substance introduced into the air that damages living things and the environment.

Sources of Air Pollution

Most air pollution comes from industry, including coal-fired power stations and mining. Incinerators used to burn waste can also release pollutants. Air pollution also comes from natural sources, such as decaying plant material, wind-blown dust, and smoke from wildfires.

Global Problems

Fast Fact

Burning some kinds of waste can release dangerous pollutants, such as the metals mercury and lead. These metals are harmful to people's health.

Some forms of air pollution create global problems by changing the amounts of gases in the atmosphere. Today, **global warming** due to increased greenhouse gases, particularly **carbon dioxide**, threatens all habitats and life-forms. In the past 150 years, human use of fossil fuels has released large amounts of carbon dioxide. Carbon dioxide levels are now 30 percent higher than at any other time in the last 650,000 years.

Smoke from wildfires can cause severe air pollution in cities.

Fast Fact
The trucks at Long Beach Port produce 10 percent of the diesel exhaust emissions in the coastal area of Los Angeles.

Trucks are responsible for much of the air pollution at Long Beach Port.

CASE STUDY

Air Pollution in Long Beach

Long Beach is in Los Angeles County, California. It has a busy port, which is located near residential areas. Many children in schools near the port suffer from breathing disorders due to air pollution.

Pollution from Long Beach Port

Ships, trucks, trains, and cargo-handling equipment at Long Beach port emit large amounts of pollutants. These include nitrogen oxides and particles from diesel engine exhaust fumes. Nitrogen oxides are gases that contribute to **smog** formation. The particles released by diesel engine exhausts are harmful to people's health.

Cleaning Up Trucks

Authorities in Long Beach have now introduced a Clean Trucks Program. It aims to reduce air pollution from trucks in the port by more than 80 percent by 2012. Trucks that do not meet emissions standards by 2012 will be banned.

Toward a Sustainable Future: Understanding and Reducing Air Pollution

Scientific research is important to understanding which pollutants in the air cause damage. For example, very fine particles in the air, called ultrafine particles, were once believed to be harmless. However, in 2008, these particles were found to increase people's risk of heart disease.

Learning About Pollution

Ongoing research is needed to learn about new kinds of pollution, including pollution from newly developed chemicals. Research can also identify chemicals that may react with other chemicals in the atmosphere to become pollutants.

International Cooperation

Reducing pollutants in the air requires international cooperation. In the atmosphere, pollution from one place can drift to others. To prevent this, countries need to agree on ways to reduce air pollution globally.

Smog forms when sunlight reacts with a mixture of smoke and sulfur in the air.

Fast Fact
The word *smog* is a combination of the words *smoke* and *fog*.

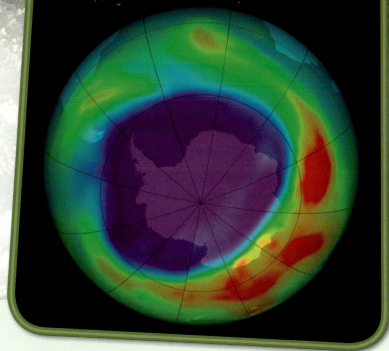

The purple color in this illustration shows the area over Antarctica where the ozone layer is thinnest.

Sources of Emissions that Damage the Ozone Layer

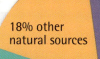

2% volcanic eruptions

18% other natural sources

80% human activity

Emissions from human activity cause the most damage to the ozone layer.

CASE STUDY

The Montreal Protocol

Reducing air pollution requires international cooperation. The Montreal Protocol is a good example of how international cooperation can protect our environment. This international agreement was established to help protect the ozone layer.

CFCs and the Ozone Layer

Chlorofluorocarbons (CFCs) were common in spray cans until 1987. When they are released into the atmosphere, CFCs reduce the ozone layer. The ozone layer filters out dangerous rays from the Sun that can cause health problems, such as skin cancers.

The Montreal Protocol and CFCs

The Montreal Protocol banned the use of CFCs and other air pollutants in 1987. This has reduced the release of CFCs and helped the ozone layer to begin recovering. The process of recovery is hard to predict. However, if all ozone-damaging chemicals are reduced, the ozone layer could recover in most areas by about 2050.

Fast Fact
International Ozone Day is celebrated each year on September 16. This is the day the Montreal Protocol was signed.

Litter

Litter is waste that has not been disposed of properly. Its negative impacts include visual pollution and physical danger to people and wildlife.

Types of Litter

Litter can be any solid or liquid waste, debris, or trash from households or businesses. This can include cigarette butts, paper, food, abandoned vehicle parts, and garden clippings. Litter along highways now accounts for almost half of all litter in many **developed countries**, such as the United States, Ireland, and Australia.

The Problem of Litter

Litter makes public places less attractive. It can block drains, and some litter, such as broken glass, can be dangerous. One of the biggest problems with litter occurs when it washes into waterways. Litter in waterways can kill plants and animals. Litter also reduces water quality, making it unsuitable for drinking and as habitat for plants and animals.

Fast Fact
One of the most common items found in litter is cigarette butts.

Litter that is dumped beside highways is a problem in many countries.

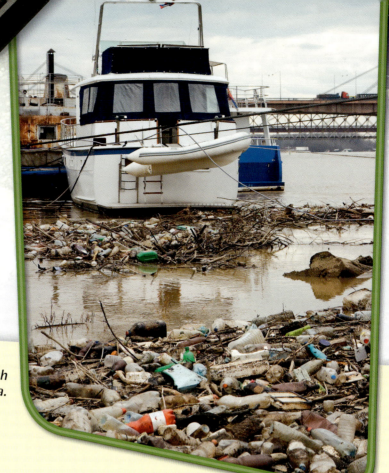

Litter that is dropped on land can wash into rivers and be carried out to sea.

CASE STUDY

The Great Pacific Garbage Patch

In the Pacific Ocean, about 3 million tons (3.2 million tonnes) of trash has collected to become a formation known as the Great Pacific Garbage Patch. Some of this trash came from ships. However, more than 80 percent of it began on land.

Litter in the Sea

Litter dropped on land can wash into drains and out to sea. Ocean currents take this litter to areas where it accumulates. The Great Pacific Garbage Patch is held together by rotating currents northeast of Hawaii.

Plastic Litter

Plastic litter lasts for a long time in water without breaking down. Large quantities of plastic litter block sunlight to the ocean. This prevents **photosynthesis** by **plankton**, which die because they cannot produce food. Plankton are the main food supply for marine animals. Therefore, the litter blocking the sunlight damages the whole ecosystem.

Toward a Sustainable Future: Cleaning Up

Systems to dispose of trash need to be put in place to keep the environment clean.

Recycling Trash

More and more trash is being **recycled** today. Recycling is developing as an industry because people are beginning to see trash, such as empty soda cans, as a worthwhile resource. However, not everything can be recycled to make new products.

Cleanup Campaigns

Much of the work to highlight problems is being done by volunteer-based organizations. Cleanup campaigns, such as Clean Up the World, have become popular all over the world.

Businesses are now being required to take responsibility for their products for the whole life of the items. This means businesses must consider the environmental effects of a product from its making through to its disposal.

Fast Fact
In 2002, it was estimated that 23 percent of all deaths are associated with environmental problems.

Many types of waste, including plastic drink bottles, can be recycled.

Ian Kiernan was one of many people who took part in Clean Up Australia Day in 2009.

CASE STUDY
Clean Up Australia Campaign

The first official Clean Up Australia Day was held in 1990. It was organized by Ian Kiernan, a solo yachtsman, to help clean up litter in Sydney Harbour.

Shocked by Pollution

In 1987, Kiernan was shocked by the pollution he saw when he took part in a solo yacht race around the world. He kept coming across polluted areas, such as the Sargasso Sea in the northern Atlantic Ocean. It was covered with litter, including old shoes, plastic buckets, disposable diapers, toothpaste tubes, and plastic bags.

Cleanup Events

Once back in Sydney, Kiernan organized Clean Up Sydney Harbour Day. About 40,000 volunteers removed plastics, glass, cigarette butts, and rusted cars from the harbor. Then, in 1990, nearly 300,000 people participated in the first Clean Up Australia Day. Today, it is one of the largest organized events in the world.

Fast Fact
Clean Up the World began in 1993 with the support of the United Nations Environment Programme.

Water Pollution

Water pollution is the presence of unwanted substances in **groundwater**, rivers, lakes, and oceans. It can cause disease and death in people and damage to natural environments. Human activities that pollute water include industry and agriculture. Untreated human waste can also sometimes reach waterways.

Oil Transportation

Oil that is transported in pipelines and tankers is a major source of water pollution. Leaks and accidents during transportation release large volumes of oil into waters every year.

The Effects of Oil Spills

The effects of oil spills depend on the type of oil and how much is spilled. However, all oil spills are dangerous for living things. Oil in water smothers plants and animals. It is toxic if it is swallowed or absorbed by living creatures.

Fast Fact
Water pollution is a leading cause of disease worldwide. It causes more than 14,000 deaths every day.

Drinking water from a polluted well can cause disease.

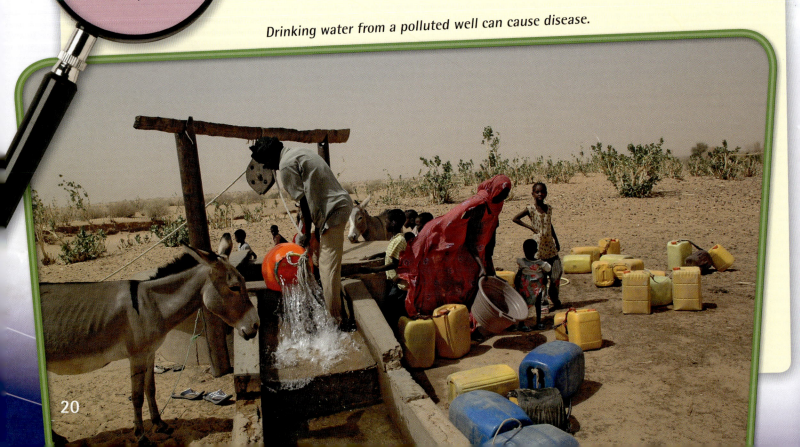

The Amoco Cadiz spilled 3,936 tons (4,000 t) of fuel, in addition to the oil.

CASE STUDY

Amoco Cadiz Oil Spill

On March 16, 1978, the oil tanker *Amoco Cadiz* ran aground off the coast of Brittany, France. About 219,432 tons (223,000 t) of oil from the ship spilled into the sea. The oil spread over 200 miles (320 kilometers) of coastline.

Loss of Marine Life

The *Amoco Cadiz* accident caused the largest loss of marine life ever from an oil spill. The accident killed 20,000 birds, and millions of snails and other sea creatures. The oil that reached the shoreline damaged all life on the shore.

The Cleanup

Strong winds and heavy seas made it difficult to clean up the oil. Clean-up activities included washing rocks and sands using high-pressure hoses, and removing the oil-covered sand. However, this process was also very damaging to life on the shores.

Toward a Sustainable Future: Cleaning Up Water Pollution

Knowing how to clean up water pollution can reduce the impact of contamination.

Cleanup Methods for Oil Spills

Cleanup methods for oil spills include the use of **chemical dispersants**. Devices called booms and skimmers are also used to contain and collect the oil. The effectiveness of the cleanup depends on the circumstances. For example, bad weather can spread the oil, making the problem larger and harder to manage.

Environmental Protection Agencies

Fast Fact

A U.K. campaign called "Bag It and Bin It" aimed to prevent disposable items being flushed down toilets. The organizers hoped this would help prevent pollution of rivers and oceans.

Many countries have environmental protection agencies to help prevent water pollution. In 1970 the United States established the Environmental Protection Agency due to public demands for a cleaner environment. The United Kingdom's Control of Pollution Act 1974 makes it illegal to pollute water. Similar agencies and laws now exist in countries across the world.

When oil spills occur close to the coast, they must be treated promptly to minimize damage.

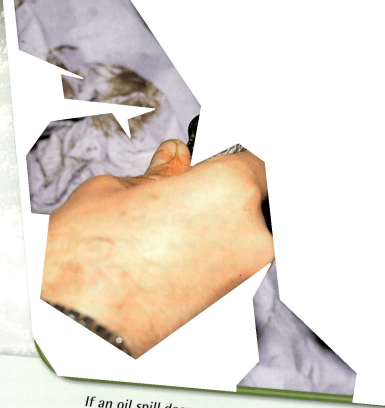

If an oil spill does reach the coast, animals that are affected need to be rescued and cleaned.

CASE STUDY

Protecting Coastal Areas from Oil Spills

Highest priority is given to protecting coastal areas from oil spills. This is because they are particularly sensitive to oil pollution.

Threats to Coasts

It is difficult to clean up oil spilled in the open sea. The spilled oil can stay afloat until it reaches coastal areas. It is sometimes possible to protect coasts by placing booms to prevent oil reaching the shore. This method works best near bays and river mouths. When oil does drift toward shores, booms can hold it back. This allows it to be treated away from plants and animals that live on the shore.

Vietnam Prepares for Oil Spills

Vietnam has developed an oil spill response plan to prepare for oil spills before they occur. The Internatioinal Union for Conservation of Nature helped develop the plan to protect the Quang Ninh and Hai Phong areas of Vietnam's coasts.

Light and Noise Pollution

Most people think of pollutants as solid materials or other physical substances. However, unwanted light and noise can also be pollutants, damaging people's health.

Light Pollution

Any unwanted light at night is light pollution. In most cities, almost all the stars are invisible at night. This is due to large numbers of lights and poorly designed light fixtures that scatter light.

Light pollution wastes energy and money. It also affects the natural environment and people's health by changing natural cycles of light and dark.

Noise Pollution

Noise pollution is unwanted or excessive sound. It can be excessive noise in a factory that damages workers' health. It can also be noise in a city that prevents people from sleeping. Noise has negative effects on people's hearing and on their overall well-being.

Fast Fact
Unwanted heat, such as hot water released from a factory into a waterway, is also a pollutant. It is known as thermal pollution.

Noise from airplanes can make living near an airport unbearable.

Fast Fact

Paris is famous for its beauty at night when buildings, such as the Eiffel Tower, are lit up. However, the lights use lots of energy, add to global warming, and cost about $22,000 a day.

Many cities in Japan are brightly lit throughout the night.

CASE STUDY
Light Pollution in Japan

Japan is one of the most light-polluted countries on Earth. Light pollution in its cities is extreme.

Problems Caused by Light Pollution

Light pollution can cause many problems. Residents in areas with excessive lighting may have to install heavy window shutters to block out the light. This kind of light pollution detracts from people's quality of life. It also costs extra money and wastes energy.

Awareness of Light Pollution

Across the world, there is a lack of awareness that light pollution can and should be reduced. The town of Bisei on the island of Honshu was the first place in Japan to recognize the problem. They introduced controls on light pollution in 1989.

25

Toward a Sustainable Future: Preventing Light and Noise Pollution

Many light and noise problems can be prevented. This can be done by reducing noise and responsibly managing human-produced light.

Controlling Light Pollution

Glare and **light trespass** need to be controlled. This can be done using well-designed light fittings. Many governments and organizations have used well-designed lighting for street lamps and stadium lighting.

It is important to keep thinking about what kinds of things cause pollution. One hundred years ago light would not have been considered pollution, but today it certainly is.

Controlling Noise Pollution

Noise pollution can be controlled by building barriers, such as freeway barriers, to screen people from noise. New types of industrial equipment can also be designed to operate more quietly.

Barriers can help reduce noise from roads.

Fast Fact

Light pollution disturbs the normal behavior patterns of many animals, including turtles. When baby turtles hatch from their eggs at night, they are attracted by light. This means they crawl inland instead of crawling safely into the ocean.

During Earth Hour in 2009, many of the lights were turned off in Las Vegas, Nevada.

CASE STUDY

National Dark-Sky Week

During National Dark-Sky Week, people all over the United States turn off unnecessary outdoor lights. This has many benefits, such as saving energy and allowing everyone see the beauty of the stars at night.

International Dark–Sky Association

The International Dark-Sky Association and the American Astronomical Society have endorsed Dark-Sky Week, which was started in 2003. More people participate every year.

Lighting and Safety

Some people believe that street lighting is needed to prevent crime. However, the International Dark-Sky Association says that no scientific studies convincingly show a relationship between lighting and crime.

ISSUE 5

What Can You Do?
Prevent Pollution

Preventing pollution is better than struggling to undo the damage it causes.

Tackle Pollution in Your Area

Step 1: Take a walk around your local area. Decide which forms of pollution are major concerns. Look for and ask about:

- visual pollution, such as litter
- light pollution from poorly designed lighting
- noise pollution from traffic or industry in residential areas

You can also check:

- how clear the air is in calm weather conditions
- how much litter and oil there is in drains and waterways

Step 2: Decide which problems should or could be tackled.

Step 3: Design a method of tackling some part of the most urgent problem. For example, you could organize a community cleanup or write to local government officials about the problem.

Fast Fact
In 2006, the city of Sao Paulo, Brazil, banned all outdoor advertising as visual pollution. Now many other cities, such as Buenos Aires in Argentina and Bergen in Norway, are following its lead.

A community cleanup is one way to tackle a litter problem.

Vu Thi Hong Hanh has written an environmental memoir about pollution in Hanoi, Vietnam.

Environmental Memoirs

The health of people and communities depends on the health of their environment. Recording personal stories or memoirs about environmental damage helps others understand the effects of these events.

The Environmental Memoirs Project

The Environmental Memoirs Project is a series of memoirs from people living in environments that are being **degraded**. These memoirs were collected and published online.

Write a Memoir

After exploring the memoirs online, collect information and write one yourself. It could be about something you or someone you know has experienced. There may be a serious pollution problem in your community. Or the problem could be small, such as litter on your street, or noise that bothers people at night.

Step 1: Write about your experiences or interview someone in your community about an environmental issue they have experienced in their lifetime.

Step 2: Document their story in the form of a memoir.

Step 3: Send your story to a local newspaper or elsewhere to get it published.

Well, I hope you now see that if you take up my challenge your world will be a better place. There are many ways to work toward a sustainable future. Imagine a world with:

- a sustainable ecological footprint
- places of natural heritage protected for the future
- no more environmental pollution
- less greenhouse gas in the air, reducing global warming
- zero waste and efficient use of resources
- a secure food supply for all

This is what you can achieve if you work together with my natural systems.

We must work together to live sustainably. That will mean a better environment and a better life for all living things on Earth, now and in the future.

Websites

For further information on pollution, visit the following websites.

- Environmental Kids Club www.epa.gov/kids/
- World Environment Day www.unep.org/wed/2009/english/
- Climate Kids http://climate.nasa.gov/kids/index.cfm
- International Dark-Sky Association www.darksky.org

Glossary

agricultural runoff
Unused nutrients from fertilizers that flow off farmland with rain or irrigation water.

algae
Living things that are found in water and make food using the energy from the Sun.

algal blooms
Huge, visible growths of algae.

atmosphere
The layers of gases surrounding Earth.

carbon dioxide
A colorless, odorless gas.

chemical dispersants
Chemicals that break down oil into tiny droplets.

chlorofluorocarbons (CFCs)
Gases that break down ozone in the atmosphere.

degraded
Run down or reduced to a lower quality.

developed countries
Countries with industrial development, a strong economy, and a high standard of living.

emissions
Substances released into the environment.

fossil fuels
Fuels such as oil, coal, and gas, which formed under the ground from the remains of animals and plants that lived millions of years ago.

global warming
An increase in the average temperature on Earth.

greenhouse gases
Gases that help trap heat in Earth's atmosphere.

groundwater
Water found below the surface of the land.

light trespass
Unwanted artificial light that shines into someone else's area.

ozone layer
A layer in the atmosphere, made up of a gas called ozone, that absorbs dangerous rays of sunlight.

photosynthesis
The process by which plants produce food from sunlight.

plankton
Tiny plants and animals that float in the sea.

pollutants
Any unwanted substances in the environment.

recycled
Made a used product or material into a new product or material.

smog
A combination of smoke and fog that is damaging to health.

sustainable
Does not use more resources than Earth can regenerate.

visual pollution
Visible substances in the environment that are unattractive and unwanted.

Index